CAMathories®

第 二 冊
dai6　yi6　chaak3
The Second Session

五 隻 馬 騮 撈 月 亮
ng5　jek3　ma5　lau1　laau4　yut6　leung6
5 Monkeys Catching the Moon

廣 東 話 詞 語 同 數 學 作 業 簿
gwong2 dung1 wa2 chi4 yu5 tung4 sou3 hok6 jok3 yip6 bou2
Cantonese Vocabulary and Mathematics Workbook

作者: Author: Kit Cheung, Ph.D. (Cantab)
編輯: Editor: Winnie Chan
插圖繪畫師: Illustrators: Dylan Parks, Katherine Mao

CAMathories Company 書 籍
CAMathories Company　syu1 jik6
CAMathories™ Books

CAMathories Company 民 間 故 事 數 學 系 列
CAMathories Company　man4 gaan1 gu3 si6 sou3 hok6 hai6 lit6
CAMathories™ Folktale Mathematics™ Series

支 持 多 元 化 同 埋 包 容 性，為 一 個 世 界 服 務
ji1 chi4 do1 yun4 fa3 tung4 maai4 baau1 yung4 sing3, wai6 yat1 go3 sai3 gaai3 fuk6 mou6
Supporting Diversity and Inclusion for One World

適 合 3 至 5 歲 嘅 課 程
sik1 hap6 saam1 Ji3 ng5 seui3 ge3 fo3 ching4
Curriculum for 3-5 years old

系 列 1: 由 1 到 5 計 數 同 背 誦
hai6 lit6 1: yau4 1 dou3 ng5 gai3 sou3 tung4 bui3 jung6
Series 1: Count and Recite 1 to 5

(有 英 國、墨 西 哥 同 中 國 民 間 故 事)
(yau5 ying1 gwok3, mak6 sai1 go1 tung4 jung1 gwok3 man4 gaan1 gu3 si6)
(Folktales from Britain, Mexico, and China)

CAMathories Company 出 版
CAMathories Company　cheut1 baan2
Published by CAMathories Company

美 國
mei5 gwok3
55 Union Place, #238, Summit, NJ 07901, U.S.A.
USA
www.camathories.com

ISBN: 978-1-962028-26-4

總 編 輯: Kit Cheung, Ph.D. (劍 橋)
jung2 pin1 chap1: Kit Cheung, Ph.D.　(gim3 kiu4)
Editor-in-Chief: Kit Cheung, Ph.D. (Cantab)

編 輯: Winnie Chan and Monika Gupta
pin1 chap1: Monika Gupta
Editor: Monika Gupta

排 版 同 校 對: Rajesh Kumar, OMNI Publication
paai4 baan2 tung4 gaau3 deui3: Rajesh Kumar, OMNI Publication
Typesetter and Proofreader: Rajesh Kumar, OMNI Publication

TABLE OF CONTENTS

目　　　　　錄

muk6　　　luk6

A. COUNTING

數

Monkey

馬　　　騮

ma5　　　lau4

One　　monkey　　has　　　two arms,　　　two hands,

一　隻　馬　騮　有　兩　隻　手　臂，兩　隻　手，

yat1　jek3　ma5　lau4　yau5　leung5　jek3　sau2　　bei3　leung5　jeg3　sau2,

two　　　legs,　　two　　feet,

兩　條　腿，兩　隻　腳，

leung5　tiu4　teui2,　leung5　jek3　geuk3,

two　　　ears,　　　two　　　eyes,

兩　隻　耳　仔，兩　隻　眼　睛，

leung5　jek3　yi5　jai2,　leung5　jek3　ngaan5　jing1,

one　　　nose,　one　　mouth,

一　個　鼻，一　個　口，

yat1　go3　bei6,　yat1　go3　hau2,

and　　　　one　　tail.

同　埋　一　條　尾。

tung4　maai4　yat1　tiu4　mei5　。

A1. How many monkeys are there in total?

呢　度　總　共　有　幾　多　隻　馬　騮?

nei1　dok6　jung2　gung6　yau5　gei2　do1　jek3　ma5　lau4?
(here)　　(in total)　　(has)　(how many)　(units of)　(monkeys)

Answer :__________________________

答　　案

daap3　on3

A2. How many monkey hands are there in total?

呢　度　總　共　有　幾　多　隻　馬　騮　手?

nei1　dok6　jung2　gung6　yau5　gei2　do1　jek3　ma5　lau4　sau2?
(here)　(in total)　(has)　(how many)　(units of)　(monkeys)　(hands)

Answer :__________________________

答　　案

daap3　on3

A3. How many monkey feet are there in total?

呢　度　總　共　有　幾　多　隻　馬騮　腳？

nei1　dok6　jung2　gung6　yau5　gei2　do1　jek3.　ma5　lau4　geuk3?

(here)　(in total)　(has)　(how many)　(units of)　(monkeys)　(feet)

Answer: ______________________

答　　案

daap3　on3

A4. How many monkey tails are there in total?

呢　度　總　共　有　幾　多　條　馬　騮　尾？

nei1　dok6　jung2　gung6　yau5　gei2　do1　tiu4　ma5　lau4　mei5?

(here)　(in total)　(has)　(how many)　(units of)　(monkeys)　(tails)

Answer: ______________________

答　　案

daap3　　on3

B. FAMILY RELATIONS
家庭關係
ga1　　ting4　　gwaan1　hai6

Monkey's Family (Siblings)

馬　騮　嘅　家　庭　(兄　　弟　姊　妹)
ma5　lau4　geei3　ga1　ting4　(hing1　dai6　ji2　mui6)

The oldest monkey

最 老 嘅 馬 騮

jeui3　lou5　geei3　ma5　lau4

The youngest monkey

最 細 (年 紀) 嘅 馬 騮

jeui3　sai3　(nin4　gei2)　geei3　ma5　lau4

Brothers and Sisters

兄 弟 姊 妹

hing1　dai6　ji2　mui6

Brothers (both the elder and the younger ones)

哥 哥 弟 弟

go4　go1　dai4　dai6

The eldest brother

大 哥 / 大 哥 哥 / 大 佬

daai6 go1 / daai6 go4 go1 / daai6 lou2

The second eldest brother

二 哥 / 二 哥 哥 / 二 佬

yi6 go1 / yi6 go4 go1 / yi6 lou2

The third eldest brother

三　　哥 / 　三　　哥哥 / 　三　　佬

saam1　go1 / 　saam1　go4　go1 / 　saam1　lou2

The little / young brother

弟　　弟 / 　細　　佬

dai4　dai6 / 　sai3　lou2

The second youngest brother

二　細　佬

yi6　sai3　lou2

The third youngest brother

三　　細　佬

Saam1　sai3　lou2

The youngest brother out of all brothers

細　　細　佬

sai3　sai3　lou2

Sisters (both the elder and the younger ones)

姐　姐　妹　妹

je4　je1　mui4　mui2

The eldest sister

大　家　姐

daai6　ga1　je1

The second elder sister

二　家　姐

yi6　ga1　je1

The third elder sister

三　家　姐

saam1　ga1　je1

Young sister

妹　　妹 / 細　　妹

mui4　mui2 / sai3　mui2

The second youngest sister

二　　妹 / 二　　妹　　妹

yi6　mui2 /　yi6　mui4　mui2

The third youngest sister

三　　妹 / 三　　妹　　妹

saam1　mui1 / saam1　mui4　mui2

The youngest sister out of all sisters

細　　妹 / 細　　妹　　妹（細　　細　　妹）

sai3　mui2　/ sai3　mui4　mui2　(sai3　sai3　mui2)

C. VOCABULARY

詞　語

chi4　　yu5

C.1. COLORS OF NATURE AND THE FOREST

大　自　然　同　森　林　嘅　顔　色

daai6　ji6　yin4　tung4　sam1　lam4　ge3　ngaan4　sik1

Rainbow　　　　color

彩　　虹　　顔　　色

paai3　hung4　ngaan4　sik1

Seven　　　　colors

七　　種　　顔　　色

chat1　jung2　ngaan4　sik1

Red 紅　色 **hung4　sik1**		**Orange** 橙　色 **chaang2　sik1**	
Yellow 黃　色 **wong4　sik1**		**Green** 綠　色 Luk6　sik1	
Light green 青　色 cheng1　sik1		**Blue** 藍　色 Laam4　sik1	
Purple 紫　色 ji2　sik1			

Different   colors

唐 同 嘅 顏 色

m4 tung4 ge3 ngaan4 sik1

A   blue   sky

一 個 藍 色 嘅 天 空

yat1 go3 laam4 sik1 ge3 tin1 hung1

One (unit)   tree

一 喬 樹

yat1 po1 syu6

Brown   tree   trunk

啡　色　樹　幹

fe1　sik1　syu6　gon3

Green   tree   leaves

綠　色　樹　葉

luk6　sik1　syu6　yip6

A   white   cloud

一　舊　白　色　雲

yat1　gau6　baak6　sik1　wan4

A   dark

一　個　黑　媽　媽

yat1　go3　hak1　ma1　ma1

dark   dark

(黑　矇　矇/ 黑　孖　孖)

(hak1　mang1　mang1/hak1　ma1　ma1)

night

嘅　夜　晚

ge3　ye6　maan5

A   yellow   moon

一　個　黃　色　月　亮

yat1　go3　wong4　sik1　yut6　leung6

C2. ANIMALS

動　物

dung6　mat6

Elephants

象/　大　象　/大　笨　象

jeung6/　daai6　jeung6　/daai6　ban6　jeung6

Leopards

豹

paau3

Panda

熊　貓

hung4　maau1

Red   Panda

紅　熊　貓

hung4　hung4　maau1

C. 3. FOUR SEASONS

四　　季
sei3　gwai3

Spring

春　　天 (= sky)
cheun1　tin1

Summer

夏　　天
ha6　　tin1

Autumn (Fall)

秋　　天
chau1　tin1

Winter

冬　　天
dung1　　tin1

D. MEASURE WORDS

量　詞

leung6　chi4

1.　(One) unit

(1)　個

yat1　go3

Usually, for round objects, including things in nature, such as food and fruits we use this measure word: "個 go3". If the object is round but very small, we will use the word "粒 lap1".

通　常　圓　形　嘅　物　體，包　括　大

tung1　seung4　yun4　ying4　ge3　mat6　tai2,　baau1　kut3　daai6

自　然　嘅　嘢，食　物　同　生　果　都　係

Ji6　yin4　ge3　ye5,　sik6　mat6　tung4　saang1　gwo2　dou1　hai6

用　呢　一　個　量　詞："個"。

yung6　ni1　yat1　go3　leung6　chi4:　"go3"。

如　果　個　圓　形　物　體　好　細，

yu4　gwo2　go3　yun4　ying4　mat6　tai2　hou2　sai3,

我　哋　就　會　用　個　"粒"　字。

ngo5　dei6　jau6　wui5　yung6　go3　lap1　ji6。

D1. NATURE

大　自　然

daai6　ji6　yin4

(One)　　Sun

一　個　太　陽

yat1　go3　taal3　yeung4

(One)　　Moon

一　個　月　亮

yat1　go3　yut6　leung6

(One)　　Earth

一　個　地　球

yat1　go3　dei6　kau4

D2. FRUITS

生　　果

saang1　go2

One　　　(big)　watermelon

一　個　(大)　西　瓜

yat1　go3　daai6　sai1　gwa1

One　　　pineapple

一　個　菠　蘿

yat1　go3　bo1　lo4

One　　　honeydew　melon

一　個　蜜　　瓜

Yat1　go3　mat6　　gwa1

One　　　pear

一　個　啤　梨

Yat1　go3　be1　lei2

One   apple

一　個　蘋　果

Yat1　go3　ping4　gwo2

One   orange

一　個　橙

yat1　go3　chaang2

One   peach

一　個　桃

yat1　go3　tou2

Small things,   usually   round-shape   or

細　嘅　嘢,　通　常　係　圓　形　或　者

sai3　ge3　ye5,　tung1　seung4　hai6　yun4　ying4　waak6　je2

oval-shape,

橢　圓　形

to5　yun4　ying4

we   use   "粒 lap1".

我　哋　會　用　一　粒。

ngo5　dei6　wui5　yung6　yat1　lap1 。

One   **seed**

一　粒　種　子

yat1　lap1　jung2　ji2

One   **raisin**

一　粒　葡　萄　干

yat1　lap1　pou4　tou4　gon1

One   **grape**

一　粒　葡　提　子

Yat1　lap1　pou4　tai4　ji2

One   **candy**

一　粒　糖

Yat1　lap1　tong2

D3. FOOD

食　　物

Some	breads	are	round
有 啲	麵 包	係	圓 形 嘅,

yau5　di1　min6　baau1　hai6　yun4　ying4　ge3,

we	also		will	use	"個 go3",	for example,
我	哋 都		會	用	"個",	例 如,

ngo5　dei6　dou1　wui5　yung6　"go3",　lai6　yu4

char siu	bao (Barbecue pork bun)	Pineapple bao (pineapple bun).
叉 燒	包,	菠 蘿 包。

cha1　siu1　baau1,　　　　bo1　lo4　baau1.

But	if	the	bread
但	係 如 果	個	麵 包

daan6　hai6　yu4　gwo2　go3　min6　baau1

has been   **by people**   **cut into**   **becoming**   **pieces or slices**

已 經 俾 人 切 咗 變 成 一 塊 塊,

yi5 ging1 bei2 yan4 chit3 jo2 bin3 sing4 yat1 faai3 faai3,

then   **we**   **call it**   **a slice of**   **bread**

咁 我 哋 就 叫 一 塊 麵 包。

gam2 ngo5 dei6 jau6 giu3 yat1 faai3 min6 baau1。

If   **it is**   **a**   **cake.**

如 果 係 一 個 蛋 糕。

yu4 gwo2 hai6 yat1 go3 daan6 gou1。

the   **cake**   **has been**   **by people**   **cut**   **into**   **several pieces,**

個 蛋 糕 已 經 俾 人 切 咗 幾 件,

go3 daan6 gou1 yi5 ging1 bei2 yan4 chit3 jo2 gei2 gin6,

then   **we**   **call it**   **one piece/slice of**   **cake.**

咁 我 哋 就 叫 一 件 蛋 糕。

gam2 ngo5 dei6 jau6 giu3 yat1 gin6 daan6 gou1。

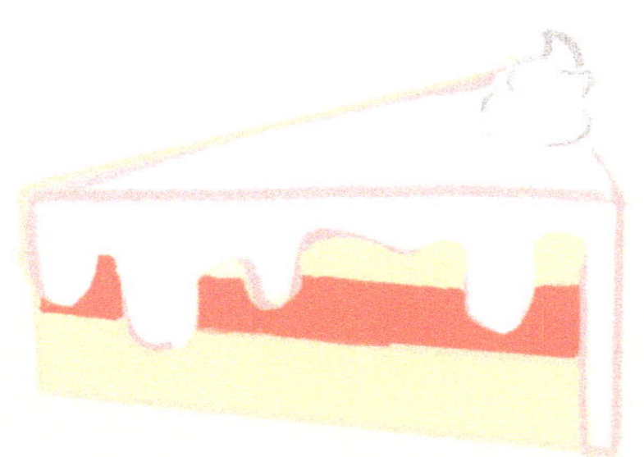

D4. PEOPLE

人　　類

yan4　　leui6

We, human beings, use the measure word "個 go3", one (個go3) human being.

我　哋　叫　人　類　做　一　個　人。

ngo5　dei6　giu3　yan4　leui6　jou6　yat1　go3　yan4。

One　　　　man

一　　個　　男　　人

yat1　go3　naam4　yan2

One　　　　woman

一　　個　　女　　人

yat1　go3　　neui5　yan2

One　　　　child

一　　個　　細　　路　　仔

yat1　go3　sai3　lou6　jai2

One   boy

一　個　男　仔

yat1 go3   naam4 jai2

One   girl

一　個　女　仔

yat1 go3   neui5 jai2

One   baby

一　個　B　B

yat1 go3   bi4 bi1

D5. ANIMALS

動　物

dung6　mat6

Usually for animals, we use the measure word, "隻 jek3"

通　常　我　哋　會　叫　一　隻　動　物，

tung1　seung4　ngo5　dei6　wui2　giu3　yat1　jek3　dung6　mat6，

for example,　one　monkey,　one　bird,　one　cow.

例　如　一　隻　馬　騮，　一　隻　雀　仔，　一　隻　牛。

lai6　yu4　yat1　jek3　ma5　lau1，　yat1　jek3　jeuk3　jai2，　yat1　jek3　ngau4。

Sometimes fruits which is long in shape, we will use

有　時　形　狀　長　嘅　水　果，　我　哋　都　會

yau5　si4　ying4　jong6　cheung4　ge3　seui2　gwo2，　ngo5　dei6　dou1　wui5

隻 as well,　For example,　one　banana.

用　一　隻，例　如　一　隻　香　蕉。

yung6　yat1　jek3，　lai6　yu4　yat1　jek3　heung1　jiu1。

Footnotes:

備　注

bei6　jyu3

Some people would use "隻" (jek3) to　count　oranges　and　apples,

有　啲　人　會　用　隻　嚟　數　橙　同　蘋　果，

yau5　di1　yan4　wui5　yung6　jek3　lei4　sou2　chaang2　tung4　ping4　gwo2，

but we never use
但 係 我 哋 從 來 都 唔 會 用
daan6 hai6 ngo5 dei6 chung4 loi4 dou1 m4 wui5 yung6

"隻 jek3", to describe big round fruits
隻 嚟 形 容 大 嘅 圓 形 水 果,
jek3 lai4 ying4 yung4 daai6 ge3 yun4 ying4 seui2 gwo2,

like watermelons
例 如 西 瓜
lai6 yu4 sai1 gwa1

or honey melons and pineapple
或 者 蜜 瓜, 同 菠 蘿。
waak6 je2 mat6 gwa1, tung4 bo1 lo4。

One bear
一 隻 熊
yat1 jek3 hung4

One cat
一 隻 貓
yat1 jek3 maau1

One rabbit
一 隻 兔 仔
yat1 jek3 tou3 jai2

One deer
一 隻 鹿
yat1 jek3 luk2

One monkey
一 隻 馬騮
yat1 jek3 ma5 lau4

One frog
一 隻 青 蛙
yat1 jek3 ching1 wa1

One cow
一 隻 牛
yat1 jek3 ngau4

One donkey
一 隻 驢
yat1 jek3 lou4

One fox
一 隻 狐狸
yat1 jek3 wu4 lei2

One squirrel
一 隻 松 鼠
yat1 jek3 chung4 syu2

One bird
一 隻 雀 仔
yat1 jek3 jeuk2 jai2

D6. KITCHEN   UTENSILS

廚　房　用　具

chyu4　fong2　yung6　geui6

Some   kitchen   utensils   for example,　a spoon,   a fork,   and

有　啲　廚　房　用　具　例　如　匙　羹,　叉,　同

yau5　di1　chyu2　fong2　yung6　geui6　lai6　yu4　　chi4　gang1,　cha1　tung4

a bowl   we also use the measure word "隻jek3".

碗,   我　哋　都　會　用　"隻"。

wun2,   ngo5　dei6　dou1　wui2　yung6　"jek3"。

Three   spoons

三　隻　匙　羹

saam1　jek3　chi4　gang1

Two   forks

兩　隻　叉

leung5　jek3　cha1

One   Bowl

一　隻　碗

yat1　jek3　wun2

E. WORDS FOR ORDER/SEQUENCE

形　容　先　後　次　序　嘅　生　字／詞　語

ying4　yung4　sin1　hau6　chi3　jeui6　ge3　saang1　ji6 / chi4　yu5

First

首　先

sau2　sin1

Following

跟　住

gan1　jyu6

Then

然　後

yin4　hau6

And

同　埋

tung4　maai4

Also

仲　有

jung6　yau5

After that

之　後

ji1　hau6

Final/Very last

最　後

jeui3　hau6

Water

水

seui2

Sky

天

tin1

Mountain

山

saan1

Moon

月　亮

yut6　leung6

Daddy

爸　爸

ba4　ba1

Mummy

媽　媽

ma4　ma1

Elder brother

哥　哥

go4　go1

Elder sister

姐　姐

je4　je1

Younger brother

弟　弟

dai4　dai6

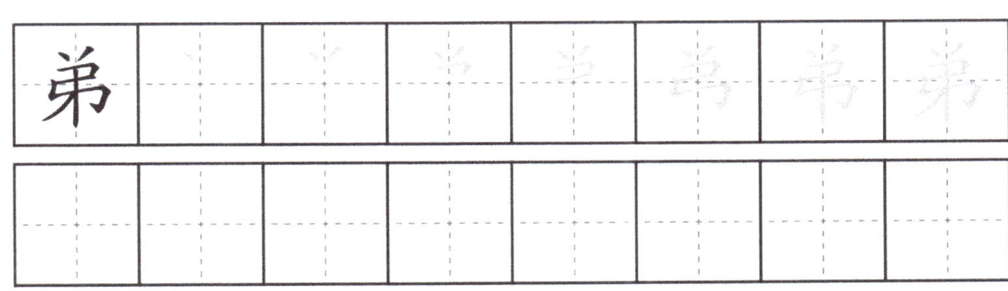

Younger sister

妹　妹

mui4　mui2

1) Make a sentence with one of the measure words: "一個yat1 go3", " 一隻yat1 jek3", or "一 粒yat1lap1".

用　其　中　一　個　量　詞:　"一　個",　"一　　隻",　或

yung6　kei4　jung1　yat1　go3　leung4　chi4,　"yat1　go3",　" yat1　jek3",　waak6

者　"一　　粒",　做　一　句　句　子。

je2　"yat1　　lap1"　jou6　yat1　geui3　geui3　ji2。

2) Make a sentence with any words of sequence: "first (首先 sau2 sin1)", "following 跟住 gan1jyu6" , "then 然後yin4 hau6" , "and 同 埋 tung4 maai4", or "after that之 後 ji1hau6".

用　其　中　一　個　形　容　先　後　次　序　嘅

yung6　kei4　jung1　yat1　go3　ying4　yung4　sin1　hau6　chi3　jeui6　ge3

生　字:　"首　先",　"跟　住",　"然　　後",　"同　　埋",

saang1　ji6:　"sau2　sin1",　"gan1　jyu6",　"yin4　hau6",　tung4　maai4",

或　者　　“之　後”，　做　一　　句　　句　　子。
waak6　je2　　"ji1　hau6"　jou6　yat1　geui3　geui3　ji2。

3) Make a sentence with any words of siblings (elder brother, elder sister, younger brother or younger sister).

用　其　中　一　個　兄　弟　姊　妹　詞　語
yung6　kei4　jung1　yat1　go3　hing1　dai6　ji2　mui6　chi4　yu5

(哥　哥，　姐　姐，　弟　弟，　妹　　妹)　做　一　　句　　句　子。
(go4　go1,　je4　je1,　dai4　dai6,　mui4　mui2)　jou6　yat1　geui3　geui3　ji2。

H. CHILDREN'S SONGS
兒　歌
yi4　　　go1

Music　　　　　and　　　　lyrics

音　樂　同　埋　歌　詞

yam1　ngok6　tung4　maai4　go1　chi4

Song

歌

go1

1. A Bird Falling into Water　falling into water,

有　隻　雀　仔　　跌　落　水

yau5　jek3　jeuk3　jai　　　dit3　lok6　seui2

E flat major:

Have　a　　bird　Falling　into　water

5　　6　　5　　4　　3　　4　　5

有　隻　雀　仔　跌　落　水,

yau5　jek3　jeuk3　jai　dit3　lok6　seui2,

Falling into water,

2　　3　　4,

跌　落　水

dit3　lok6　seui2

Falling into water,

3　　4　　5,

跌　落　水

dit3　lok6　seui2

Have a bird falling into water,
5   6   5   4   3   4   5,

有 隻 雀 仔 跌 落 水

yau5 jek3 jeuk3 jai dit3 lok6 seui2

Let water flushed away.
2   5   3   1

被 水 沖 去。

bei6 seui2 chung1 heui3。

FURTHER   CHALLENGE

更　　多　　挑　　戰

gang3　do1　tiu1　jin3

Song

歌

go1

2. Shiny Shiny Moon, brighten the front yard

月　　光　　光,　　照　　地　　堂

yut6　gwong1　gwong1,　jiu3　dei6　tong4

Eb Major

Moon　shiny　shiny,

　12　　53　　　23

月　　光　　光,

yut6　gwong1　gwong1,

Reflecting on the front yard,

　12　　76　　　5

照　　地　　堂,

jiu3　dei6　tong4,

Baby   Har,   you,   be good,   sleep on bed.   (*蝦=shrimp milkname)
3   2   1   5   3 2   1 2   7 6   5
蝦*   仔   你   乖   乖   瞓   落   床。
ha1   jai2   nei5   gwaai1   gwaai1   fan3   lok6   chong4。

Next morning,   mommy   needs to rush to   plant rice seeds,
5   5 3   1   2   1   2   1 3   2   1
聽   朝   阿   媽   要   趕   插   秧   咯，
ting1 jiu1   a3   ma1   yiu3   gon2   chaap3   yeung1   lo3， (adverb)

Grandpa   watches ox he   goes to the mountain,   (adverb)
1   5   2   5   1   1 2   5 3   3 2   312532 312765
阿   爺   睇   牛   佢   上   山   崗，   啊⋯⋯啊
a3   ye4   tai2   ngau4   keui5   seung5   saan1   gong1， a1……a3

Baby   Har You   grow fast,   tall   and   big,
3   2   1   1   53   2   6   11
蝦   仔   你   快   高   長   大   咯，
ha1   jai2   nei5   faai3   gou1   jeung2   daai6   lo3， (adverb)

Helping   Grandpa to   watch   ox   goat .   (adverb)
3   23   2   6   1   23   3 6   5.   172765 172532
幫   手   阿   爺   去   睇   牛   羊。   啊⋯⋯啊
bong1 sau2   a3   ye4   heui3   tai2   ngau4   yeung4。 a1   a3

Moon   shiny shiny,
12   53   23

月   光   光，

yut6 gwong1 gwong1,

Reflecting on the front yard,
12   7 6   5

照   地   堂，

jiu3 dei6 tong4,

Baby Har, you,   be good,   sleep on bed.
3   2   1   5   3 2   1 2 7 6 5

蝦 仔 你 乖 乖 瞓 落 床。

ha1 jai2 nei5 gwaai1 gwaai1 fan3 lok6 chong4。

Next morning father needs to catch fish shrimp (adverb),
5   53   2   1   2   1   76   5   23   1

聽 朝 阿 爸 要 捕 魚 蝦 咯，

Ting1 jiu1 a3 ba4 yiu3 bou6 yu2 ha1 lo3,

Grandma knits nest needs to knit till sky bright, (adverb)
1   5   2   1   1   2   1   53   32   312532   312765

阿 嫲 織 網 要 織 到 天 光 ， 啊…… 啊……

a3 ma4 jik1 mong5 yiu3 jik1 dou3 tin1 gwong1, a3…… a3….

Baby Har, you, quickly be tall grow big (adverb),
3 2 1 1 53 2 6 11

蝦 仔 你 快 高 長 大 咯 ，

ha1 jai2 nei5 faai3 gou1 jeung2 daai6 lo3,

Row a boat cast a net is even better /good.
6 53 5 5 3 5 35 2 3 5

划 艇 撒 網 就 更 在 行 。

wa1 teng5 saat3 mong5 jau6 gang3 joi6 hong4。

Moon shiny shiny
12 5 3 2 3

月 光 光 ，

yut6 gwong1 gwong1

Reflecting on front (earth) yard,
12 7 6 5

照 地 堂 ，

jiu3 dei6 tong4,

Year 30th night (At the end of the year)
5 27 1

年 卅 晚 ，

nin4 sa1 maan5 ,

Pick　betel　nut
61　　26　　5
摘　　檳　　榔，
jaak5 ban1 long4,

Five　grain　full　harvest　pack　full　the warehouse　(adverb),
12　　53　　3　　32　　3　　12　　3　　21
五　　穀　　豐　　收　　堆　　滿　　倉　　囉，
ng5　guk1　fung1　sau1　deui1　　mun5　chong1　　lo1,

Old　old　young　young　happy　so　so,　　(adverb)
1　　1　　6　　6　　23　　36　　5　　172765　1615321
老　老　嫩　嫩　　喜　洋　洋,　　啊……　啊……
lou5 lou5 nyun6 nyun6　　hei2　yeung4　yeung4,　　a1…..a1……

Baby　Har　you　quickly　close　up　Eyes　　(adverb),
3　　2　　1　　1　　5 3　　2　　3 5　　1　　1
蝦　仔　你　快　啲　瞇　埋　眼　囉，
ha1　jai2　nei5　faai3　di1　mei1　maai4　ngaan5　lo1 ,

(dawn)
One　sleep　gets　to　big　sky　bright.　(adverb)
3　　23　　2　　2 3　　12　　53　　32　　372765 1725321
一　覺　瞓　到　大　天　光。　啊……
yat1　gaau3　fan3　dou3　daai6　tin1　gwong1。　a1……

I. ART AND CRAFTS

Drawing

畫　畫

waak6　wa2

5 Monkeys

五　隻　馬　騮

ng5　jek3　ma5　lau

1 bunch of bananas

一　梳　香　蕉

yat1　so1　heung1　jiu1

J. FINAL MATHART DRAWING CHALLENGE

最　後　數　學　藝　術　挑　戰

jeui3　hau6　sou3　hok6　ngai6　seut6　tiu1　jin3

Draw a picture of the forest using any shapes you learn from Workbook 1 (goldilocks and the three bears). Hint: triangle, square or circle.

用　你　喺　第　一　本　作　業　簿 (高　迪　樂　絲

yung6　nei5　hai2　dai6　yat1　bun2　jok3　yip6　bou2　(gou1　dik6　lok6　si1

同　三　隻　熊)　學　到　嘅　任　何　形　狀　畫

tung4　saam1　jek3　hung4)　hok6　dou2　ge3　yam6　ho4　ying4　jong6　waak6

一　幅　森　林　嘅　圖　畫。提　示：三　角　形,

yat1　fuk1　sam1　lam4　ge3　tou4　wa2。tai4　si6：saam1　gok3　ying4,

正　方　形　或　圓　形。

jing3　fong1　ying4　waak6　yun4　ying4。

Answers

答　　　案

daap3　on3

A1.

There　are　in total　(having)　5　(unit)　monkeys.

呢　度　總　共　有　五　隻　馬　騮

nei1　dok6　jung2　gung6　yau5　ng5　jek3　ma5　lau4

A2.

There　are　in total　(having)　10　(unit)　monkey　hands

呢　度　總　共　有　十　隻　馬　騮　手

nei1　dok6　jung2　gung6　yau5　sap6　jek3　ma5　lau4　sau2

A3.

There　are　in total　(having)　10　(unit)　monkeys　feet

呢　度　總　共　有　十　隻　馬　騮　腳

nei1　dok6　jung2　gung6　yau5　sap6　jek3　ma5　lau4　geuk3

A4.

There　are　in total　(having)　5　(unit)　monkeys　tails

呢　度　總　共　有　五　條　馬　騮　尾

nei1　dok6　jung2　gung6　yau5　sap6　jek3　ma5　lau4　mei5

有　關　作　者
yau5　gwaan1　jok3　je2
About the Author

Kit Cheung　博　士（劍　橋）
Kit Cheung　bok3　si6　(gim3　kiu4)

Kit Cheung　博　士　喺　唔　同　嘅　地　方　生　活、學
Kit Cheung　bok3　si6　hai2　ng4　tung4　ge3　dei6　fong1　sang1　wut6,　hok6

習　同　工　作　過,　由　中　國　嘅　香　港　去
jaap6　tung4　gung1　jok3　gwo3,　yau4　jung1　gwok3　ge3　heung1　gong2　heui3

新　加　坡,　到　蘇　格　蘭　愉　快　熱　鬧　嘅
san1　ga3　po1,　dou3　sou1　gaak3　laan4　yu4　faai3　yit6　naau6　ge3

皇　家　里　爾　（Royal Mile）,　再　到　英　格　蘭
wong4　ga1　leui5　yi5　(Royal Mile),　joi3　dou3　ying1　gaak3　laan4

如　童　話　般　嘅　劍　橋,　跟　住　去　咗　美
yu4　tung4　wa2　bun1　ge3　gim3　kiu4,　gan1　jyu6　heui3　jo2　mei5

國　巴　爾　的　摩,　世　界　上　螃　蟹　餅　嘅
gwok3　ba1　yi5　dik1　mo1,　sai3　gaai3　seung6　pong4　haai5　beng2　ge3

首　都,　最　後　安　家　喺　花　園　之　州　嘅
sau2　dou1,　jeui3　hau6　on1　ga1　hai2　fa1　yun2　ji1　jau1　ge3

新　澤　西　州。
san1　jaak6　sai1　jau1.

Kit Cheung　博　士　鍾　意　喺　亞　洲、歐　洲　同　澳
Cheung　bok3　si6　jung1　yi3　hai2　a3　jau1,　au1　jau1　tung4　ou3

大　利　亞　洲　不　同　嘅　地　方　遊　歷,　同
daai6　lei6　a3　jau1　bat1　tung4　ge3　dei6　fong1　yau4　lik6,　tung4

唔　同　嘅　人　傾　偈,　了　解　唔　同　嘅　文
ng4　tung4　ge3　yan4　king1　gai2,　liu5　gaai2　ng4　tung4　ge3　man4

化,　對　唔　同　區　域　嘅　民　間　故　事　深
fa3,　deui3　ng4　tung4　keui1　wik6　ge3　man4　gaan1　gu3　si6　sam1

感　興　趣。
gam2　hing3　cheui3.

Kit Cheung 博 士 係 英 國 外 交、 聯 邦 同
Cheung bok3 si6 hai6 ying1 gwok3 ngoi6 gaau1, lyun4 bong1 tung4

發 展 辦 公 室 頒 發 嘅 Chevening 獎 學
faat3 jin2 baan6 gung1 sat1 baan1 faat3 ge3 Chevening jeung2 hok

金 同 劍 橋 大 學 外 國 留 學 信
gam1 tung4 gim3 kiu4 daai6 hok6 ngoi6 gwok3 lau4 hok6 seun3

託 頒 發 嘅 劍 橋 海 外 信 託 獎
tok3 baan1 faat3 ge3 gim3 kiu4 hoi2 ngoi6 seun3 tok3 jeung2

學 金 嘅 獲 獎 者, 同 時 取 得 咗
hok6 gam1 ge3 wok6 jeung2 je2, tung4 si4 cheui2 dak1 jo2

劍 橋 大 學 人 文 與 社 會 科 學 學
gim3 kiu4 daai6 hok6 yan4 man4 yu5 se5 wui2 fo1 hok6 hok6

院 嘅 博 士 學 位。
yun2 ge3 bok3 si6 hok6 wai2.

Cheung 博 士 係 CAMathories Company 嘅 共
Cheung bok3 si6 hai6 CAMathories Company ge3 gung6

同 創 辦 人 同 埋 首 席 執 行 官。
tung4 chong3 baan6 yan4 tung4 maai4 sau2 jik6 jap1 hang4 gun1.

About the Author
Kit Cheung, Ph.D. (Cantab)

Dr. Kit Cheung has lived, studied, and worked in different places, from the fun yet hustle and bustle cities of Hong Kong (China) and Singapore, to the awesome Royal Mile in Edinburgh, Scotland, then to fairytale-like Cambridge in England, and later to Baltimore, the crab cake capital of the world, before settling in the garden state, New Jersey, of the U.S.A. Dr. Cheung likes to travel to different places in Asia, Europe, and Australasia, to meet different people, to learn about different cultures, and is fascinated by different folk stories from different areas.

Dr. Cheung is a recipient of the Chevening Scholarship (awarded by the British Foreign, Commonwealth and Development Office, U.K.) and Cambridge Overseas Trust (by the University of Cambridge, U.K.), and has obtained a Ph.D. degree from the School of Humanities and Social Sciences at the University of Cambridge. Dr. Cheung is a co-founder and the Chief Executive Officer of CAMathories Company.

This work book belongs to :

呢　本　作　業　簿　係　屬　於:

ni1　bun2　jok3　yip6　bou2　hai6　suk6　yu1

Date

日　　期

yat6　kei4

Cutting page

Copyright © 2026 CAMathories Company. U.S.A.